Real Steam in Colour
Volume 1

York

CW01669694

FRONT COVER: Class B16/3 61464 makes a spirited departure from York with a Scarborough to Leeds express in May 1959. Sister locomotive 61456 waits for the road on the avoiding line with an up freight. 61464 was constructed at Darlington Works in December 1923 as LNER 1381 and survived until September 1963, 61456 being withdrawn in August 1960.
RETINA IIIC, 50mm f2 Xenon, Kodachrome I, 1/250 f2.4

REAR COVER: Class J72 68736 stands in the carriage sidings at the east side of York station in June 1960. With sister engine 68723, it was repainted in the original NER light green livery during the course of general repairs at Darlington Works in May 1960. The engines carried both the BR emblem and the NER Coat of Arms on the tank sides. 68736 was based at York until 1961 when it joined 68723 at Newcastle to perform carriage pilot duties.
LEICA M2, 50mm f1.4 Summilux, Kodachrome I, 1/125 f4.5

ABOVE: Class V2 60974 departs from the south end of York station with an up express on a sunny morning in May 1959. Built as LNER 3686 at Darlington Works in August 1943, it was renumbered as LNER 974 in May 1946 and survived without rebuilding as BR 60974 until withdrawal in December 1963. The classic lines of the arched station roof and York South roundhouse can be seen in the picture.
RETINA IIIC, 50mm f2 Xenon, Kodachrome I, 1/250 f2.4

Designed and Typeset by Lomax Publicity, Falmouth
Printed by BDC Printing Services, Derby
Bound by Hammond Bindery, Wakefield
Colour separations by Scantec Lithographics, Falmouth
ISBN 0-9516690-0-1

Published by
Norseman Publications Ltd, 27 Vyvyan Street, CAMBORNE, Cornwall, TR14 8BQ, England.
NORSEMAN

Introduction

With its magnificent Minster and City Walls, York is not only one of England's finest cities, but is also a railway centre of great significance. Situated mid-way between the English and Scottish capitals, and at the meeting point of routes from the Midlands, Lancashire and the South West, it was always destined to play a major role in the growth of Britain's railways.

It was put 'On the Map' by Railway King, George Hudson, whose favourite architect George Townsend Andrews designed many of the railway buildings, some of which have survived to this day. Many of the early railway companies merged to form the North Eastern Railway in 1854. This company, in 1877, opened a massive new through station which was topped by five magnificent arches to support its awe inspiring roof, the largest of these arches standing 48ft high and having a span of 81ft.

In the days before the introduction of restaurant cars, expresses stopped in the station to allow passengers to snatch a hasty meal at the refreshment rooms whilst a fresh locomotive was being provided from one of York's motive power depots. In 1877, the North

ABOVE: Class A3 60075 ST. FRUSQUIN heads north from York past Leeman Road mpd with an express for Newcastle in August 1959. ST. FRUSQUIN was one of twenty A1 Class Pacifics constructed by the North British Locomotive Company in 1924. Built as LNER 2574 (Works No. 23112), it was rebuilt as Class A3 in 1943 and was later fitted with a Kylchap double chimney in August 1959. This particular locomotive was never fitted with German type smoke deflectors and survived until January 1964 having run 2,077,000 miles in traffic. The flower beds visible in the foreground were part of a 1930s scheme introduced by the Chief Area Engineer, John Miller, as part of a determined effort to brighten the lineside. This was achieved by laying lawns, planting shrubs, and surrounding the plots with concrete edging made by the LNER at York.

RETINA IIIC, 50mm f2 Xenon, Kodachrome I, 1/250 f2.4

Eastern Railway also opened a new depot at York North at a cost of £36,700. The original depot consisted of three adjoining roundhouses, two of which were converted at a cost of £984,000 to form the National Railway Museum.

The museum, opened by HRH the Duke of Edinburgh on September 28th 1975, houses exhibits from the former historical collection of the British Transport Commission and British Railways Board, mainly from the now-closed museums at Clapham and York. It provides an educational insight to Britain's railway heritage, although nothing can recreate the feeling for the true working railway as depicted in these pages.

Acknowledgements are made to the PR and PO of British Railways, North Eastern Region for granting photographic facilities. Thanks also go to the RCTS whose volumes of detailed information were invaluable in compiling this work. Lastly, I am indebted to Ernest Sanderson for the details given in his book "Railway Memories No I - YORK."

Keith R Pirt Chesterfield

ABOVE: On a lovely late afternoon in May 1959 Class A2/3 60517 OCEAN SWELL departs from one of York station's western platforms with a down express. 60517 was built at Doncaster Works in October 1946 (Works No. 2008). A Gateshead based locomotive for many years, it was reallocated to Darlington in June 1961 and was shuttled between that depot, Heaton and Gateshead before being withdrawn from Gateshead in January 1964. The station at York survived for over 70 years with few major alterations until 1951 when some of the world's most impressive mechanical signal-boxes gave way to just one control centre situated in the station. This meant that the wooden slotted signals which had guarded the station approaches for many decades were replaced by neat and efficient colour light signals.

RETINA IIIC, 50mm f2 Xenon, Kodachrome I, 1/250 f2.4

North of York Station

ABOVE: Class V2 60930 emits a smoke laden exhaust and passes the lines of locomotives outside Leeman Road mpd whilst working a northbound extra in August 1959. Built at Doncaster in July 1941 as LNER 3657 (Works No. 1923), it was renumbered 930 in January 1947 and survived as BR 60930 until September 1962. On the NER section the V2s showed their ability to run particularly well on the level racing ground across the Plain of York. Speeds of up to 85 mph were achieved and sustained averages of 80 mph with loads of 400 tons were run on a road little easier then level. Such running compared well with performances by contemporary Class 7 locomotives on other lines.
RETINA IIIC, 50mm f2 Xenon, Kodachrome I, 1/250 f2

BELOW: The best known of the A3 Class, 60103 FLYING SCOTSMAN, rests in Leeman Road up loop with a southbound empty stock and parcels train in August 1959. Built at Doncaster Works in February 1923 (Works No. 1564) as LNER 1472, it was renumbered to the familiar 4472 in February 1924. After further renumbering to 103 in May 1946, it was rebuilt from Class A1 to Class A3 in January 1947. In January 1959 a Kylchap double chimney was fitted and German type smoke deflectors were added in December 1961. After completing some 2,076,000 miles in revenue earning traffic, it was withdrawn by BR in January 1963 and was bought by Alan Pegler for preservation.
RETINA IIIC, 50mm f2 Xenon, Kodachrome I, 1/250 f2.4

BELOW: Whilst FLYING SCOTSMAN steals all of the limelight today, there was a period when the names of other locomotives of the same class were as often on our lips, and with equal admiration. Here A3 60050 PERSIMMON heads a Newcastle express past Leeman Road mpd sidings in May 1959. Originally built at Doncaster Works as Class A1 2549 (Works No. 1605), PERSIMMON was rebuilt to Class A3 in December 1943. A Kylchap double chimney was added in April 1959 as can be seen in this photograph, but it wasn't until October 1961 that German type smoke deflectors were added. The fitting of a Kylchap double chimney was to prevent the problem of steam and smoke drifting back along the boiler and obstructing the engineman's view of the road ahead. The first experiments in this direction were made in 1937 to 2751 HUMORIST. At first a lipped-top chimney was fitted without deflector plates but the drifting back of the exhaust was so severe that a plain chimney and two small deflector plates were soon substituted, and the engine ran with these until 1947 when it received the standard deflector plates as fitted to the new Thompson Class A1s.
RETINA IIIC, 50mm f2 Xenon, Kodachrome I, 1/250 f2.4

ABOVE: Class A4 60016 SILVER KING passes Leeman Road sidings with a down express on a sunny afternoon in May 1959. Built at Doncaster Works as LNER 2511 in November 1935 (Works No. 1821), SILVER KING was the third member of the A4 Class to be built. It was outshopped in the silver and grey livery for working the 'Silver Jubilee' express. By the mid-1950's it was one of eight A4 Class Pacifics allocated to Gateshead, many of which were rarely seen in the clean condition of their Kings Cross or Haymarket classmates. 60016 left its East Coast main line duties in November 1963 when it was reallocated to Aberdeen Ferryhill where it was used on the 3 hour services to Glasgow until final withdrawal came in March 1965.

RETINA IIIC, 50m f2 Xenon, Kodachrome I, 1/250 f2.4

BELOW: Class A2/2 60501 COCK O' THE NORTH heads an express for Newcastle past Leeman Road carriage sidings in May 1959. Originally constructed in May 1934 as LNER Class P2 2-8-2 2001 it was subsequently rebuilt as an A2/2 Pacific in September 1944. As a class, the P2s did all that was required in terms of load haulage on the gradients, but fell short of a desirable standard of availability and gave trouble with overheating. Whilst these problems could be contained in peacetime, it made the engines almost impossible to operate in wartime. However, it can be said that after rebuilding as Pacifics, they were unable to match the uphill performance of the original 2-8-2s.

Having spent much of its early life on the tortuous gradients between Edinburgh and Aberdeen, COCK O' THE NORTH came south in 1949 when the new Peppercorn Class A2 Pacifics took over its duties. 60501 saw out its days at York until it was withdrawn from traffic in February 1960.

RETINA IIIC, 50 mm f2 Xenon, Kodachrome I, 1/250 f2.4

BELOW: Class A4 60030 GOLDEN FLEECE heads the down 'Flying Scotsman' past Leeman Road carriage sidings in October 1959. Visible in the photograph is the distinctive Haymarket based headboard with its light blue background. Constructed at Doncaster as LNER 4495 GREAT SNIPE (Works No. 1861) in August 1937, it ran with this name for about one month before it was renamed GOLDEN FLEECE. When built, it carried the garter blue livery with red wheels for working the 'Coronation' and 'West Riding' trains. When introduced, the high speed East Coast main line services were the fastest in Britain. However, this gave rise to another major problem - how to stop them. In fact, when travelling at 95 mph, it was impossible to stop the trains between distant and home signals. This problem was presented to the Westinghouse Brake & Signal Co., who, after considerable testing, developed the Quick Service Application valve which drastically improved the braking of these new high-speed trains.

RETINA IIIC, 50mm f2 Xenon, Kodachrome I, 1/250 f2.4

ABOVE: Class A4 60027 MERLIN approaches York with the up 'Elizabethan' whilst J72 68677 shunts in the Clifton carriage sidings in May 1959. At this time, it is interesting to note that the ageing J71 and J72 pilots continued to be housed in the York South roundhouse even though it was literally falling down around them. Eventually it became unsafe to enter the building until the roof was removed and the walls remained standing. From that time it continued to be used by the pilots until it was demolished in 1963.

Always a Scottish engine, 60027 was allocated to Haymarket from new until February 1962, when it was transferred to St Rollox for use on the Glasgow to Aberdeen 3 hour services. After subsequent reallocation to St Margarets in September 1964, it was withdrawn from service in September 1965.

RETINA IIIC, 50 mm f2 Xenon, Kodachrome I, 1/250 f2.4

BELOW: Class A3 60091 CAPTAIN CUTTLE runs light engine on the main line from York station to York North mpd. Clearly visible in the background, York Minster is undergoing repair. The date is August 1959. Built at Doncaster in September 1928 as LNER 2745 (Works No. 1695), it carried the number 91 from October 1946 until it received the number 60091 upon the incorporation of British Railways in April 1948. As can be seen in this picture, it had received the Kylchap double chimney in March 1959, although the German trough type smoke deflectors were not fitted until October 1961. Transferred from Heaton to Gateshead in June 1962, it was withdrawn from the latter depot in October 1964 after running 1,562,315 miles in traffic.

RETINA IIIC, 50mm f2 Xenon Kodachrome I, 1/250 f2.4

BELOW: Class J27 65894 rests between shunting duties in the permanent way yard at Clifton in October 1965. Constructed by the LNER at Darlington Works to NER Class P3 design as No 2392, it was later renumbered 5894 in March 1946. For many years it was allocated to York North mpd, but its final allocation was to Sunderland mpd where it joined the remaining members of the class working the colliery branches before withdrawal in September 1967. Unlike the remainder of the class, 65894 was purchased for preservation by the North Eastern Locomotive Preservation Group and after restoration to North Eastern Railway livery as NER 2392 it has worked on the North Yorkshire Moors Railway as well as returning to York for a brief spell on exhibition in the National Railway Museum.
LEICA M3, f1.4 Summilux, Kodachrome II, 1/250 f4

ABOVE: Class J72 68677, resplendent in clean unlined BR black livery, rests outside the cleaning shed at Clifton carriage sidings in August 1959. Constructed at Darlington Works in December 1898 as NER 1746, it was allocated from new to York mpd where it was employed on freight shunting duties. The NER number was retained at the 1923 grouping until it was renumbered as LNER 8677 in May 1946. When taken into BR stock in 1948 it became 68677 and at the same time it was fitted with vacuum brakes enabling it to work passenger stock. By the time of its withdrawal in October 1961, it had spent all of its 62 plus years at York.

RETINA IIIC, 50mm f2 Xenon, Kodachrome I, 1/125 f4

ABOVE: Class A1 Pacifics 60155 BORDERER and 60146 PEREGRINE arrive on the main line opposite York North shed prior to reversing into the shed yard. The clean lined green livery stands out in the clear late afternoon sun in October 1964. The A1s were the last of the LNER Pacifics to be built, the main series being constructed in 1948/9. At that time they were principally used on the heavier workings on the East Coast main line, at a time when the older Gresley Pacifics were operating under something of a cloud.

Developed by Peppercorn from Thompson's 1945 rebuild of the pioneer Gresley Pacific GREAT NORTHERN, 60146 was built at Darlington in April 1949 (Works No. 2065) whilst 60155 was built at Doncaster in September 1949. Both locomotives were allocated to York from where they were withdrawn in October 1965. One feature to be noted on 60155 was that Timken roller bearings were fitted to the driving axles.

LEICA M3, 50mm f1.4 Summilux, Kodachrome II, 1/250 f3.5

York North mpd

BELOW: Class A2/2 60501 COCK O' THE NORTH prepares to reverse out of York North shed yard prior to working north with an express to Newcastle in May 1959. Controversy about this class of engines will continue, but it must be stated that without a shadow of doubt, the strength of the class as 2-8-2s, at low speeds on steep gradients can never be questioned. In everyday, regular service and in all weathers, there was no type of British express locomotive that could take away a 550-600 ton train from an uphill start and climb the bank with equal certainty.

RETINA IIIC, 50mm f2 Xenon, Kodachrome I, 1/125 f4

ABOVE: Class A3 60060 THE TETRARCH of Gateshead mpd is depicted at the eastern end of York North shed yard in May 1959. Rarely photographed in colour, THE TETRARCH was built at Doncaster in April 1925 as LNER Class A1 2559 (Works No. 1615). Rebuilt as Class A3 in January 1942, it became LNER 60 in June 1946 and received the number 60060 upon Nationalisation in 1948.

Whilst the locomotive was fitted with a Kylchap double chimney in March 1959, it was never fitted with the German type smoke deflectors. After amassing a mileage of 1,991,512, it was withdrawn from Gateshead mpd in September 1963.

RETINA IIIC, 50mm f2 Xenon, Kodachrome I, 1/125 f4

BELOW: A general view of York North shed yard with the towers of the Minster dominating the background. V2 60895 is in the foreground, K1 62061 and two class B1s can be seen behind. The first V2 to be built in June 1936 carried the name GREEN ARROW and subsequent members of the class became known as Green Arrows. The name itself referred to a fast overnight goods from King's Cross. In the years following Nationalisation, the V2s received the lined black mixed traffic livery although this was replaced with the express passenger green livery in the 1950s. The V2s history was somewhat mixed, but it can be claimed that they were a success in the conditions for which they were built.
LEICA M3, 90mm f2 Summicron, Kodachrome II, 1/250 f4

ABOVE: Class V2 60938 allocated to March (31B) mpd stands with other locomotives at the eastern end of York North shed yard, with its clean green livery glistening in the October 1959 sunshine. 60938 has steam to spare. In the background is another V2, while a BR Standard Class 3 is being coaled. The Mitchell mechanical coaling plant was commissioned in 1932 and operated four chutes fed from bunkers having a total capacity of 500 tons. The last steam locomotives to leave York were transferred to other depots on 25th June 1967 and the coaling plant was later demolished only with some difficulty. This was followed in June 1973 by the demolition of the nearby ferro-concrete 100,000 gallon water tank. The NER was a pioneer in the use of reinforced concrete and incorporated it in many large structures pre-1914, the water tower at York being erected in 1909.

RETINA IIIC, 50mm f2 Xenon, Kodachrome I, 1/125 f 3.5

RIGHT: Another feature of York North shed yard was the four roads of ashpits with water sumps. Some engines stand on the southerly ashpit while others stand between the pits and the main line, presumably awaiting their final trip to the scrap yard.
LEICA M3, 50mm f1.4 Summilux, Kodachrome II, 1/250 f4

BELOW: Delivered to York in 1962, this Cowans-Sheldon steam breakdown crane No DB 967159 was photographed in York North shed yard in June 1965. This machine was one of twelve 75 ton cranes ordered from Cowans-Sheldon as part of the Modernisation Plan in 1960. Previously, cranes had maximum lifting capacities of 40 to 50 tons, but with the new larger diesels weighing over 120 tons, it was thought necessary to increase the capacities so that two cranes could lift these new locomotives. To achieve this the jibs had to be big and heavy with large counterbalancing. Careful handling was therefore essential otherwise the crane would topple over, as did happen several times. However, these problems were overcome in the next build of 75 ton cranes to emerge from Cowans-Sheldon, six examples being supplied to BR in 1977 with telescopic jibs. In addition to this modification, the counterweights can be moved towards and away from the fulcrum point. Both this batch of cranes and those built in 1960, which were converted from steam to diesel in 1978, are still part of the breakdown fleet in use on BR.
LEICA M3, 50mm f1.4 Summilux, Kodachrome II, 1/250 f4

ABOVE: York North shed could be relied upon to host ex-Works engines running in from both Darlington and Doncaster Works. Here, in June 1959, Class L1 67745 was noted next to the main line. Fully repainted in lined black livery, it was in transit back to Hitchin (34D) mpd. Less successful than the B1 4-6-0s, the Thompson Class L1 2-6-4Ts suffered from axlebox and motion wear which gave rise to a distinctive clanking that enabled them to be identified before they appeared in sight. Several of the class were based on the GE lines where they were divided between Norwich, Ipswich, Bishops Stortford and Stratford.
RETINA IIIC, 50mm f2 Xenon, Kodachrome I, 1/125 f4

BELOW: V2 60922 of Heaton mpd and ex-Works in lined green livery stands in York North shed in August 1959. Having been repaired at Doncaster Works, it ran back to Plant Works later that same day. Built at Darlington Works in October 1941 as LNER 4893, it was renumbered LNER 922 in December 1946 and became BR 60922 upon Nationalisation in 1948. The V2s based at York had a poor reputation, although those based in Scotland were more popular despite the heavy banks, sharp curves and frequent stops which seemed to cause motion and axleboxes to knock and glands to blow. However, as time ran out for steam, the V2s suffered along with the rest. Even as late as 1966, a V2 with 9 bogies ran at speeds up to 82 mph with sustained running at 75 mph between Darlington and York.
RETINA IIIC, 50mm f2 Xenon, Kodachrome I, 1/125 f4

LEFT: Ex-Works LMS 8F 2-8-0 48712 stands in the yard at York North mpd in October 1965. 48712 was returning to Carnforth after repairs at Darlington. At that time in the mid-1960s, Darlington Works was also repairing a variety of LM motive power, some being allocated in the West Riding of Yorkshire. A one time LNER Class 06 locomotive, 48712 was constructed at Brighton in July 1944 for the MOD. Whilst on loan to the LNER, it carried the number 7658 before being renumbered 3107 in April 1946. Subsequently it received the number 3507 in March 1947 before transfer to the LMS in October 1947 when it became 8712. It then received the prefix 4 upon Nationalisation and survived in this guise to become one of the last survivors in its class before withdrawal from Carnforth mpd in June 1967.
LEICA M3, 50mm f1.4 Summilux, Kodachrome II, 1/250 f3.5

BELOW: WD 2-8-0 90610 has been fully repainted during a visit to Darlington Works. It stands in York North shed yard on its way back to Wakefield mpd in October 1964. Built as WD 77466 by the Vulcan Foundry in January 1944, it survived almost to the end of steam traction on BR, being withdrawn from Wakefield mpd in May 1967, before being broken up at Arnott Young, Parkgate and Rawmarsh in September 1967.
LEICA M3, 50mm F1.4 Summilux, Kodachrome II, 1/125 f5

ABOVE: Class A3 60092 FAIRWAY of Heaton mpd rests in York North shed yard before returning north with an express for Newcastle. This is an August 1959 picture. This particular A3 is notable in that it was the last member of the class to have a single chimney which it carried until it was removed during a visit to Doncaster Works in October 1959. In this picture it is seen with a non-banjo dome boiler of type 94HP. Constructed in October 1928 at Doncaster (Works No. 1700) as LNER 2746, it carried this number until October 1946 when it became LNER 92. A double Kylchap chimney and a banjo dome boiler were fitted in November 1959 and German trough type smoke deflectors were added in October 1961 and withdrawal from Gateshead mpd came in October 1964.

RETINA IIIC, 50mm f2 Xenon, Kodachrome I, 1/125 f4

RIGHT: BR Standard Class 3 2-6-2T 82027 stands by the turntable at the end of York North shed yard in August 1959. A Malton based locomotive, it was a frequent visitor to York, being used extensively on local services. Constructed at Swindon Works in January 1954, it was transferred from Malton to Nine Elms in October 1964 and after only twelve years in traffic it was withdrawn from that depot in January 1966.
RETINA IIIC, 50mm f2 Xenon, Kodachrome I, 1/125 f4

BELOW: Double chimneyed BR Standard Class 9F 92205 stands in the shed yard at York North mpd. One of five 9Fs to be allocated to York in mid-1963, 92205 was kept in clean condition prior to transfer to Wakefield in October 1966 from where it was withdrawn in June 1967. The 9F was probably the most successful BR standard type locomotive design and the last standard type to be constructed, but sadly it was destined to be very short lived. Many of the class had tragically short careers, being victims of the Modernisation Plan, and were still in excellent condition when withdrawn from service. In fact, during the course of 1965 the Western Region withdrew its entire allocation of the class with included some engines which had been built little more than five years earlier. By the end of 1966, half the class had gone.
LEICA M3, 50mm f1.4 Summilux, Kodachrome II, 1/250 f4

ABOVE: LM 6P Jubilee 45717 DAUNTLESS of Bank Hall (Liverpool) mpd arrives on York North shed after working to York with a Liverpool to Newcastle express. For many years the Liverpool to Newcastle expresses were worked by a Bank Hall 6P locomotive, usually DAUNTLESS, 45698 MARS, or unrebuilt Patriot 45517. Routed via Rochdale and the Calder Valley line, these expresses were survivors of the old Lancashire and Yorkshire Railway. A Bank Hall locomotive for much of its working life, 45717 remained there until it was withdrawn in November 1963 and after being stored at Crewe Works until December 1963, it was towed to Cowlairs Works where it was scrapped in February 1964.
RETINA IIIC, 50mm f2 Xenon, Kodachrome I, 1/250 f2.4

BELOW: LM 6P Patriot 45517 of Bank Hall mpd stands resplendent in clean BR lined green livery in York North shed yard in May 1959. Having worked a Liverpool to Newcastle express as far as York, it returned to Liverpool later that afternoon. The only Patriot to be allocated to Bank Hall at that time, it survived at that depot until it was withdrawn from traffic in June 1962. Designed by Fowler, the unrebuilt series of Patriot class locomotives were the oldest passenger engines to survive into the 1960s on the London Midland Region. Although never outstanding in service, they performed steadily throughout their lives and were often found deputising for more powerful locomotives. Despite being designated passenger locomotives, the authorities ran out of inspiration and ten of the class were never named. In their latter years they were relegated to freight duties, the survivors being placed in and out of store.

RETINA IIIC, 50mm f2 Xenon, Kodachrome I, 1/125 f4

ABOVE: Following modernisation in 1957, the remaining twin roundhouse at York North mpd remained a busy shed for more than ten years. In an evocative scene of sunlight and steam gathered around its 60 and 70ft turntables in October 1964 are V2 60886 flanked by 9F 92205 and WD 90517. For many years York North shed's top link steam turns were mainly out and back workings to Newcastle, often working in one direction with an express and returning with a braked goods in the other. In addition to these turns there were York's local workings to Harrogate, Hull, Normanton, Leeds, Pickering, Whitby and Scarborough. However, with the introduction of the 'Deltic' locomotives and accelerated timings, a York crew could work to London and back in a shift, a round trip of 376 miles which qualified them for a worthwhile mileage payment.

LEICA M3, 50mm f1.4 Summilux, Kodachrome II, 1/30 f2

ABOVE: Class J27 0-6-0 65823 awaits minor steam trimming repairs whilst V2s 60961 and 60837 flank A1 60121 SILURIAN in the background. The adjacent No 3 and No 4 roundhouses at York North were later to become the main display hall of the National Railway Museum. Use of this site for railway purposes goes back to 1875 when the North Eastern Railway Board authorised construction of a shed to hold 60 locomotives. The original shed consisted of three adjoining roundhouses in the shape of an L, each with a 45ft turntable serving 20 roads. In 1891 a 50ft turntable replaced the 45ft one in No 2 shed, and in 1911 two electrically-operated 60ft turntables were ordered, one replacing the 45ft turntable in No 3, the other being used in a new No 4 shed. In June 1990 the 60ft turntable was removed from the NRM just leaving the 70ft turntable.

LEICA M3, 50mm f1.4 Summilux, Kodachrome II, 1/30 f2

York Station

BELOW: In August 1959 Class A3 60078 NIGHT HAWK, with a Newcastle to King's Cross express, runs to a halt in platform 8 of York station under the graceful arches of the overall roof. In the early hours of 29th April 1942 much of the station was devastated by a German bombing raid. The most damaged platforms were Nos 1, 2 and 3. One of the roof columns on platform 3 was completely destroyed by a high explosive bomb and an incendiary bomb fell on the lamp room near platform 15. At that time it had been used to store 500 gallons of paraffin. Despite the devastation, traffic was moving again by the evening of the same day and within twenty four hours all running lines could be used for traffic.
RETINA IIIC, 50mm f2 Xenon, Kodachrome I, 1/250 f2.4

ABOVE: Class B1 4-6-0 61377 of Doncaster mpd makes a slow start from York station with a semi-fast to Doncaster in May 1959 whilst Class A3 60036 COLOMBO stands in platform 9 with a Newcastle to Leeds express. On the right can be seen the tracks leading to the original 1841 station. At that time it was still being used for stabling coaching stock. At that time also, part of the old station was being used as a loading bay for car carrier services, later known as Motorail. In 1966 all track to the old station was lifted, the Motorail terminal moved to the 'new' station to make way for a new technical office block called Hudson House.

RETINA IIIC, 50mm f2 Xenon, Kodachrome I, 1/250 f2.4

ABOVE: In May 1959, B1 Class 4-6-0 61379 MAYFLOWER of Immingham mpd stands just south of York station with the stock of the semi-fast to Doncaster. Alongside is LMS 'Jinty' 3F 0-6-0T 47556 with two vans. Several of these engines were transferred to York in 1958 to replace the ageing J72 Class 0-6-0Ts on shunting duties and in that capacity they survived until 1961. B1 61379 was built in June 1951 by the North British Locomotive Company (Works No 26838) and survived at Immingham until withdrawal in August 1962. 61379 was the only B1 to be named after Nationalisation in July 1951 when it was given the name MAYFLOWER to signify the ties between the towns of Boston U.S.A. and Boston England. Small plaques on the cabsides were fitted to commemorate the event.
RETINA IIIC, 50mm f2 Xenon, Kodachrome I, 1/125 f4

South of York Station

BELOW: The LMS 'Crab' 2-6-0s were always frequent visitors to York and even in their original red livery and 13XXX numbers could be seen working through to Scarborough. This view of 42901 was taken as it heads south from York with a Scarborough to Leeds and Bradford train of LMS stock. It is passing a V2 2-6-2 with a flared tender and 1950 built J72 Class 0-6-0T 69016. At this time LMS 2-6-0, 0-6-0 and 4-6-0 locomotives were regular visitors to York working through to Scarborough with summer holiday trains.

RETINA IIIC, 50mm f2 Xenon, Kodachrome I, 1/250 f2.4

ABOVE: Whilst the Lowestoft-York express regularly brought East Anglian Britannia Pacifics to York, 70000 BRITANNIA and A3 60038 FIRDAUSSI pass York South mpd on a light engine working in October 1959. Both engines have been repaired and fully repainted at Doncaster Plant and are returning to Doncaster Carr mpd on a running in turn. Built at Crewe in 1951, 70000 worked on the GE until May 1963 when it was reallocated to Crewe North. After withdrawal in June 1966 BRITANNIA was stored at various locations on BR and is now preserved in working order on the Nene Valley Railway. 60038 was built at Doncaster in August 1934 as LNER 2503 FIRDAUSSI and survived as such until it was renumbered LNER 38 in October 1946. As BR 60038, only one month before this photograph was taken, FIRDAUSSI was fitted with a Kylchap double chimney which it carried until withdrawal in November 1963. By that time it had completed 1,637,492 miles in traffic and unlike many members of the class, it was spared the fitting of German type trough smoke deflectors.
RETINA IIIC, 50mm f2 Xenon, Kodachrome I, 1/250 f2.4

RIGHT: Class A2/2 60503 LORD PRESIDENT stands outside York Queen St. shed as main line standby in May 1959. Rebuilt from Gresley Class P2 2-8-2 2003 at Doncaster in December 1944 (Works No 1836), it was renumbered LNER 503 in June 1946 and 60503 in September 1948. As a rebuild from the 2-8-2, it was as liable to slip as any 4-6-2. At times the 4-6-2s could climb banks well, particularly if they could be approached at high speed, but more frequently slipping reduced the effective capacity to that of the smaller V2s. Towards the end, the survivors of this class were reduced to mainly standby duties, 60503 being withdrawn only 6 months after this picture was taken.
RETINA IIIC, 50mm f2 Xenon, Kodachrome I, 1/125 f4

BELOW: Class B16 4-6-0s 61471, 61428 and an unidentified member of the class are stored outside the roundhouse at York South mpd in 1959. In their time the class was amongst the most useful engines to be based at York. With one member of the class (LNER 925) destroyed beyond repair in the air raid of 1942, the remaining 69 members of the class were all allocated to York from 28th March 1943.
RETINA IIIC, 50mm f2 Xenon, Kodachrome I, 1/125 f4

BELOW: Class B16/3 61463 makes a resounding start from York with a Scarborough to Leeds train in May 1959. Judging from the smoke rising from the station, several other locomotives are present, an A1 being discernible on the right hand side of the picture. Built as LNER 1380 at Darlington Works it was renumbered LNER 1463 in December 1946. In November 1947 it was rebuilt under Thompson with new frames and three independent sets of Walschaerts valve gear together with a new cab, altogether a more modern design. Renumbered 61463 in May 1950, it remained at York until December 1962 when it was when it was transferred to Hull (Dairycoates) from where it was withdrawn in June 1964.
RETINA IIIC, 50mm f2 Xenon, Kodachrome I, 1/250 f2.4

BELOW: Class B16/1 4-6-0 61477 takes a northbound freight from Holgate reception sidings around the curve onto the station avoiding line in May 1959. 61477 was built at Darlington Works, under Raven, in April 1920 as NER 848. It retained this number at the grouping until December 1946 when it became LNER 1408. It should have been given the number 61408 on Nationalisation, but owing to new B1 Class construction in 1949, it was allocated the number 61477. An old class of locomotive, it was not surprising that the first member (61474) was withdrawn in February 1958, 61477 being withdrawn from York in February 1960.

RETINA IIIC, 50mm f2 Xenon, Kodachrome I, 1/250 f2.4

ABOVE: Another B16/1 Class locomotive 61412 of Neville Hill (Leeds) mpd makes a smoky departure from York with a Scarborough to Leeds train in August 1959. On the far right of the picture can be seen Queen Street shed. Formerly the works boiler shop, the shed was taken over by the Lancashire and Yorkshire Railway, but upon acquisition by the LMS it was closed in the 1930s while that company took its activities to York South sheds. After that time it was used for stabling locomotives, particularly in the 1960s when there were standbys for diesel failures. Prior to the roof falling in during 1968, it had been used for stabling officers' saloons. Shortly afterwards, the shed was demolished.
RETINA IIIC, 50mm f2 Xenon, Kodachrome I, 1/250 f2.4

RIGHT: Just over three years before the total extinction of the class, K3 61929 runs south through York station with an up freight in August 1959. This was particularly notable in that most freights used the station avoiding line. Designed by Sir Nigel Gresley at the Doncaster Works of the Great Northern Railway in 1920, the first batch of ten were notable on more than one account. The most prominent feature was the huge boiler of no less than 6ft in diameter and in this respect they were not exceeded by any other locomotive in the country. Other than the 2-6-0s of the GWR, they were the first 2-6-0s to be produced in quantity by any British railway. The locomotives also had three cylinders at a time when few engines with other than two or four cylinders were at work.
RETINA IIIC, 50mm f2 Xenon, Kodachrome I, 1/250 f2.4

BELOW: Class B16/2 61437 of York North shed takes the avoiding line at Holgate Junction with a northbound freight in May 1959. Built at Darlington Works in January 1923 as LNER 2366, it was later rebuilt by Gresley with longer frames, a new cab and two sets of outside Walschaerts valve gear; a 2:1 lever operated the inside cylinder valves. Only seven engines were rebuilt in this form before the death of Sir Nigel Gresley. Renumbered as LNER 1437 in December 1946 and becoming 61437 in January 1949, it was reallocated to Hull (Dairycoates) in December 1962 from where it was withdrawn in June 1964.
RETINA IIIC, 50mm f2 Xenon, Kodachrome I, 1/250 f2.4

ABOVE: Class A3 60074 HARVESTER of Neville Hill (Leeds) mpd approaches York station with the northbound 'North Briton': express in May 1959. Always maintained in pristine external condition by Neville Hill shed, the locomotive and train make a glorious sight in the spring sunshine. With the introduction of accelerated schedules in September 1957 it was found that the A3s were suffering from steaming troubles, particularly on the longer duties. After this time, the class was gradually fitted with Kylchap double blastpipes, a modification costing only £195 for each locomotive, the outlay being more than justified by the greatly improved reliability and consistently better performances obtained. This modification was carried out to 60074 in March 1959 and was carried until its withdrawal in April 1963, the locomotive having completed 2,170,917 miles in revenue earning traffic.

RETINA IIIC, 50mm f2 Xenon, Kodachrome I, 1/250 f2.4

BELOW: Class V2 60983 of King's Cross mpd (Top Shed) departs from York with that most famous of summer holiday trains 'The Scarborough Flyer' express to London King's Cross in June 1960. The ultimate member of the V2 class, 60983 was constructed at Darlington Works in July 1944 as LNER 3695 (Works No 1936) and was renumbered to LNER 983 in May 1946. Becoming 60983 after Nationalisation in March 1950, it retained its monobloc cylinders until it was withdrawn in September 1962.

LEICA M3 50mm f1.4 Summilux, Kodachrome I, 1/250 f2.4

BELOW: One of the more common locomotive classes at York, LNER Thompson B1 61334 departs with the York to Bournemouth through train in May 1959. Comprising mainly SR green stock with red and green coaches interspersed, this train was routed via the GC main line, and after reversal at Sheffield Victoria it ran via Woodford Halse, Banbury, Oxford and Reading, where it traversed the west curve towards Basingstoke joining the SR main line to Bournemouth. 61334 would be used as far as Sheffield where another B1 would take over the working as far as Oxford. Here an SR Lord Nelson or Bulleid Light Pacific would replace the B1 for the final leg to Bournemouth – an altogether interesting journey. For many years a Sheffield Darnall locomotive, 61334, stayed in the Sheffield area being allocated to both Millhouses and Canklow before transfer to Gorton in 1963 from where it was withdrawn in December of that year. After a period in store at Annesley, it was finally broken up at W. Rigley, Bullwell Forest Wagon Works in August 1964.

RETINA IIIC, 50mm f2 Xenon, Kodachrome I, 1/250 f2.4

ABOVE: Another Class B1 4-6-0 61382 passes under Holgate Road bridge with a heavy down freight in May 1959. Carrying the lined black livery, the tender bears the BR lion and wheel emblem. Built by the North British Locomotive Company in September 1951 (Works No. 26831), 61382 remained a Darlington based engine until September 1964 when it was reallocated to Ardsley from where it was withdrawn only three months later. Probably the most successful of Edward Thompson's designs, the B1 was the LNER equivalent of the LMS Stanier Class 5. Very much influenced by the LMS product, a total of 410 B1 4-6-0s were eventually built, 59 being named, many after species of antelope.
RETINA IIIC, 50mm f2 Xenon, Kodachrome I, 1/250 2.4

BELOW: LMS Jubilee 4-6-0 6P 45572 EIRE leaves York with the morning express to Bristol in May 1959. A Bristol (Fishponds) locomotive, 45572 and her crew face a 212 mile journey ahead, a taxing duty for both engine and footplatemen on this daily turn. A regular performer on this train, EIRE survived at Bristol until October 1961 when it was transferred to Shrewsbury. Two years later 45572 was displaced again to Willesden and after only two months in traffic was placed in store until scrapping some five months later. 1964 saw sixty-six Jubilees withdrawn from traffic, followed by another thirty-three in 1965 and seven in 1966. 1967 was the last year for the class, all eight survivors being based on the NER in the Leeds, Bradford and Wakefield ares, the last of their major duties being to haul extras over the Settle and Carlisle route, the final member, 45562 ALBERTA, being withdrawn in November.

RETINA IIIC, 50mm f2 Xenon Kodachrome I, 1/250 f2.4

ABOVE: Un-named LMS Patriot 4-6-0 6P 45517 of Bank Hall mpd, Liverpool, passes under Holgate Road bridge with the late afternoon express from Newcastle to Liverpool in May 1959. Less than twelve months later, the first of the class, 45500 PATRIOT had been withdrawn. By the end of 1961 eleven of the class had been withdrawn and by the end of 1962 they were decimated by a further twenty five withdrawals including 45517 which was withdrawn in June. Unfortunately, none of the class were preserved.

RETINA IIIC, 50mm f2 Xenon Kodachrome I, 1/250 f2.4

BELOW: Class A3 60056 CENTENARY, still with single chimney, departs from York with a Newcastle to King's Cross express in May 1959. Based at Grantham mpd, 60056 was that depot's final A3 to retain its single chimney. This is a rare picture at that date, the whole class having been fitted with double chimneys by November of that year.

Built at Doncaster in February 1925 as Class A1 Pacific LNER 2555 (Works No. 1611), CENTENARY was rebuilt as class A3 in August 1944, before being renumbered LNER 56 in July 1946 and 60056 in May 1949. Only two months after this picture was taken, 60056 was fitted with a Kylchap double chimney and was subsequently fitted with German type smoke deflectors in August 1961. Remaining a Grantham based locomotive for the rest of its career, CENTENARY survived until May 1963 having completed 1,896,016 miles in traffic.

RETINA IIIC, 50mm f2 Xenon, Kodachrome I, 1/250 f2.4

ABOVE: Class B1 61337 approaches Holgate Road bridge with a local for Sheffield in May 1959. York B1 Class 4-6-0s were regularly used on this all stations working to Sheffield Midland. A York based locomotive for most of its life, 61337 remained at York until June 1967 when York lost its allocation of steam locomotives. It was then transferred to Low Moor mpd (Bradford) when along with 61030 and 61306, it shared the distinction of being one of the last three members of the class to survive. However, all three were withdrawn from Low Moor in September 1967, and of these, only 61306 survives in preservation.

It is of interest to note that after many years of use on the continent, the 24 hour timetable was introduced as an experiment at York in November 1961. Since the 24 hour timetable has now been in use for many years, it is obvious that the experiment must have been a success, although the NER'S idea of what constituted a 'Principal Express' was somewhat erratic.

RETINA IIIC, 50mm f2 Xenon, Kodachrome I, 1/250 f2.4

BELOW: Class D20 4-4-0 62387 draws to a halt at Holgate Bridge platform with an RCTS special from Leeds on 23rd June 1957. Although regular main line engines in NER and LNER days, the last surviving D20s in the York area were based at Selby and worked local trains to Doncaster and York. Unusually 62387 was fitted with a group standard tender instead of the normal NER railed tender. The platforms at Holgate Bridge were regularly used for special events on the Knavesmire, although until August 1939, they had been mainly used for traffic to York Races. This visit by the RCTS special is believed to be the last time the platform was used before it was demolished in 1964.

VOIGTLANDER VITO B, 50mm f3.5 Color-Skopar, Kodachrome I, 1/100 f4.5

ABOVE: Class A4 60026 MILES BEEVOR heads the up 'Scotch Goods' past the lineside allotments south of Holgate Bridge platform in October 1959. Normally a King's Cross V2 working, this up train was unusually graced by an A4, although the down working was a regular A4 diagram. 60026 was built at Doncaster in February 1937 as LNER 4485 KESTREL (Works No. 1850). It was first renumbered 587 in April 1946 before it became 26 in May 1946. After renaming to MILES BEEVOR in November 1947, it eventually carried the number 60026 from September 1949. It was subsequently rebuilt with a Kylchap double chimney in August 1957 and after reallocation to Ferryhill (Aberdeen) in December 1964 it was withdrawn in December 1965. After several periods in store at Ferryhill, Perth and Crewe Works, it was one of four members of the class to be broken up at Hughes Bolckows, North Blyth.

RETINA IIIC, 50mm f2 Xenon, Kodachrome I, 1/250 f2.4

ABOVE: Class A2/3 60511 AIRBORNE passes Dringhouses Yard with a Newcastle to King's Cross express in October 1959. A BR 350hp diesel shunter can be seen on the right with another being visible in the down reception sidings on the left. 60511 was built at Doncaster (Works No. 2002) as LNER 511 in July 1946. Renumbered 60511 in April 1948, it spent most of its working life allocated to Heaton mpd and was transferred to Tweedmouth in October 1961 where it survived until October 1962 when it was withdrawn.
RETINA IIIC, 50mm f2 Xenon, Kodachrome I, 1/250 f2.4